The Ultimate Science Trivia Book

KJ Smith

Table of Contents

Every effort was made to ensure all the questions in this book were accurate at the time of publishing.

Part I. Science History

Question 1.
Q: Who is known as the father of modern physics?
A: Isaac Newton

Question 2.
Q: In what year was the structure of DNA first described, and by whom?
A: 1953, by James Watson and Francis Crick

Question 3.
Q: What ancient civilization is credited with the development of the first known written numerical system?
A: Sumerians (Mesopotamia)

Question 4.
Q: Who proposed the heliocentric model of the solar system in the 16th century?
A: Nicolaus Copernicus

Question 5.
Q: What is the name of the book published by Charles Darwin in 1859 that laid the groundwork for evolutionary biology?
A: "On the Origin of Species"

Question 6.
Q: Who discovered penicillin, and in what year?
A: Alexander Fleming, 1928

Question 7.
Q: The period known as the Scientific Revolution began in approximately what century?
A: The 16th century

Question 8.
Q: Who is credited with the invention of the telescope in the early 17th century?
A: Hans Lippershey (Galileo Galilei improved it and made significant discoveries using it)

Question 9.
Q: In what year did the first human landing on the moon occur?
A: 1969

Question 10.
Q: What ancient Greek mathematician is known as the father of geometry?
A: Euclid

Question 11.
Q: Who formulated the laws of motion and universal gravitation in the 17th century?
A: Isaac Newton

Question 12.
Q: What is the name of the first programmable computing device, invented by Charles Babbage in the 19th century?
A: The Analytical Engine

Question 13.
Q: Who discovered the electron in 1897?
A: J.J. Thomson

Question 14.
Q: The theory of relativity was developed by which physicist?
A: Albert Einstein

Question 15.
Q: What ancient scientist is credited with establishing the principles of levers and buoyancy?
A: Archimedes

Question 16.
Q: In which century did Galileo Galilei make his pioneering observations that supported the Copernican theory?
A: The 17th century

Question 17.
Q: Who is known for her work on radioactivity and won two Nobel Prizes in different scientific fields?
A: Marie Curie

Question 18.
Q: The term "cell" in biology was coined by Robert Hooke in what century?
A: The 17th century

Question 19.
Q: Who developed the first successful polio vaccine?
A: Jonas Salk

Question 20.
Q: The first artificial satellite, Sputnik, was launched by which country?
A: Soviet Union

Question 21.
Q: What physicist is known for formulating the quantum theory?
A: Max Planck

Question 22.
Q: The discovery of the circulation of blood in the human body is attributed to whom?
A: William Harvey

Question 23.
Q: Who authored the work "Principia Mathematica," a landmark in the history of science?
A: Isaac Newton

Question 24.
Q: 24. The Phlogiston theory, which was later disproved, pertained to which field of science?
A: Chemistry

Question 25.
Q: What was the name of the spacecraft that carried the first humans to the moon?
A: Apollo 11

Question 26.
Q: In what year was the periodic table created by Dmitri Mendeleev?
A: 1869

Question 27.
Q: Who is considered the father of modern chemistry?
A: Antoine Lavoisier

Question 28.
Q: The first law of thermodynamics, also known as the law of conservation of energy, was formulated in which century?
A: The 19th century

Question 29.
Q: Who discovered the planet Neptune, and in what year?
A: Johann Galle, 1846 (based on predictions by Urbain Le Verrier and John Couch Adams)

Question 30.
Q: What scientist is known for the discovery of X-rays?
A: Wilhelm Conrad Röntgen

Question 31.
Q: In what year did the Human Genome Project officially begin?
A: 1990

Question 32.
Q: Who was the first woman to win a Nobel Prize, and in what field?
A: Marie Curie, Physics

Question 33.
Q: The concept of natural selection is associated with which two scientists?
A: Charles Darwin and Alfred Russel Wallace

Question 34.
Q: What ancient physician is referred to as the father of medicine?
A: Hippocrates

Question 35.
Q: The discovery of Uranus in 1781 was made by which astronomer?
A: William Herschel

Question 36.
Q: Who is known for the development of the first successful photographic process?
A: Louis Daguerre

Question 37.
Q: In what year was Pluto reclassified as a "dwarf planet"?
A: 2006

Question 38.
Q: The first successful test of a nuclear weapon was conducted in what year?
A: 1945

Question 39.
Q: What mathematician developed the first algorithm intended to be processed by a machine?
A: Ada Lovelace

Question 40.
Q: Who discovered the principle of buoyancy, often illustrated by the story of a golden crown?
A: Archimedes

Question 41.
Q: In which century was the first mechanical calculator developed?
A: The 17th century (by Blaise Pascal)

Question 42.
Q: Who formulated the law of conservation of mass, crucial to the field of chemistry?
A: Antoine Lavoisier

Question 43.
Q: The first vaccine was developed against which disease?
A: Smallpox (by Edward Jenner)

Question 44.
Q: What is the name of the landmark experiment that demonstrated the existence of electromagnetic waves?
A: Hertz's experiments (by Heinrich Hertz)

Question 45.
Q: Who developed the germ theory of disease in the 19th century?
A: Louis Pasteur

Question 46.
Q: The first artificial heart transplant was performed in what year?
A: 1982

Question 47.
Q: Who was the first person to observe bacteria and protozoa, using a microscope he improved himself?
A: Antonie van Leeuwenhoek

Question 48.
Q: In what year did the Rosetta Stone help scholars begin to decipher Egyptian hieroglyphs?
A: The stone was discovered in 1799, but deciphering began in the early 19th century.

Question 49.
Q: What scientist is credited with establishing the classification system for living organisms?
A: Carl Linnaeus

Question 50.
Q: The discovery that the Earth's continents move and have moved over geological time scales is known as what theory?
A: Plate tectonics

Part II. Astronomy

Question 1.
What is the largest planet in our solar system?
a) Earth
b) Jupiter
c) Saturn
d) Neptune

Question 2.
Who was the first human to travel into space?
a) Neil Armstrong
b) Yuri Gagarin
c) Buzz Aldrin
d) Alan Shepard

Question 3.
What is the name of our galaxy?
a) Andromeda
b) Whirlpool
c) The Milky Way
d) Triangulum

Question 4.
What type of star is the Sun?
a) Red Giant
b) White Dwarf
c) G-type main-sequence star
d) Neutron Star

Question 5.
How many moons does Mars have?
  a) 1
  b) 2
  c) 4
  d) 6

Question 6.
What planet is known as the "Red Planet"?
  a) Mercury
  b) Venus
  c) Mars
  d) Jupiter

Question 7.
What is the brightest star in the night sky?
  a) Alpha Centauri
  b) Betelgeuse
  c) Sirius
  d) Vega

Question 8.
Which planet has the most extensive ring system in our solar system?
  a) Jupiter
  b) Saturn
  c) Uranus
  d) Neptune

Question 9.

What is the name of the first artificial satellite sent into space?

a) Explorer 1
b) Sputnik 1
c) Voyager 1
d) Hubble

Question 10.

What phenomenon occurs when the Moon passes directly between the Earth and the Sun, blocking the Sun's light?

a) Lunar Eclipse
b) Solar Eclipse
c) Transit
d) Opposition

Question 11.

Who proposed the laws of planetary motion in the early 17th century?

a) Isaac Newton
b) Galileo Galilei
c) Johannes Kepler
d) Tycho Brahe

Question 12.
What is the term for a system of two stars that orbit each other?
   a) Binary star system
   b) Double galaxy
   c) Twin nebulae
   d) Solar duo

Question 13.
What is the largest moon in the solar system?
   a) Titan
   b) Ganymede
   c) Callisto
   d) Europa

Question 14.
What dwarf planet was discovered in 2005 and is named after a Polynesian creator of humanity?
   a) Eris
   b) Makemake
   c) Haumea
   d) Ceres

Question 15.
What is the name of the telescope launched into space in 1990 to provide clear and deep images of the universe?
   a) James Webb Space Telescope
   b) Kepler Space Telescope
   c) Hubble Space Telescope
   d) Spitzer Space Telescope

Question 16.

What is the term for the point in a planet's orbit when it is closest to the Sun?

a) Aphelion
b) Apogee
c) Perihelion
d) Perigee

Question 17.

Which planet has a day longer than its year?

a) Jupiter
b) Mars
c) Venus
d) Mercury

Question 18.

What is the name given to the boundary around a black hole beyond which no light or other radiation can escape?

a) Dark zone
b) Event horizon
c) Absolute limit
d) Void edge

Question 19.
What term describes the study of the universe and its origins?
- a) Astrobiology
- b) Astrophysics
- c) Cosmology
- d) Exoplanetology

Question 20.
What is the name of the first rover to explore Mars, launched by NASA?
- a) Curiosity
- b) Opportunity
- c) Spirit
- d) Sojourner

Answers

1. b) Jupiter
2. b) Yuri Gagarin
3. c) The Milky Way
4. c) G-type main-sequence star
5. b) 2
6. c) Mars
7. c) Sirius
8. b) Saturn
9. b) Sputnik 1
10. b) Solar Eclipse
11. c) Johannes Kepler
12. a) Binary star system
13. b) Ganymede
14. c) Haumea
15. c) Hubble Space Telescope
16. c) Perihelion
17. c) Venus
18. b) Event horizon
19. c) Cosmology
20. d) Sojourner

Part III. Chemistry

Question 1.
What is the most abundant gas in the Earth's atmosphere?
  a) Oxygen
  b) Hydrogen
  c) Nitrogen
  d) Carbon dioxide

Question 2.
Who is known as the father of modern chemistry?
  a) Antoine Lavoisier
  b) Dmitri Mendeleev
  c) John Dalton
  d) Albert Einstein

Question 3.
What is the chemical symbol for gold?
  a) Au
  b) Ag
  c) Go
  d) Gd

Question 4.
Which element has the highest electrical conductivity?
- a) Copper
- b) Silver
- c) Gold
- d) Aluminum

Question 5.
What is the main component of natural gas?
- a) Methane
- b) Ethane
- c) Propane
- d) Butane

Question 6.
What is the pH level of pure water at 25°C?
- a) 7
- b) 5
- c) 8
- d) 9

Question 7.
Who created the first periodic table of elements?
- a) Albert Einstein
- b) Dmitri Mendeleev
- c) Marie Curie
- d) Isaac Newton

Question 8.

What type of bond is formed when two atoms share electrons?

a) Ionic bond
b) Covalent bond
c) Metallic bond
d) Hydrogen bond

Question 9.

Which of the following is NOT a noble gas?

a) Neon
b) Argon
c) Chlorine
d) Helium

Question 10.

What is the process called when a solid changes directly into a gas?

a) Deposition
b) Sublimation
c) Condensation
d) Evaporation

Answers

1. c) Nitrogen
2. a) Antoine Lavoisier
3. a) Au
4. b) Silver
5. a) Methane
6. a) 7
7. b) Dmitri Mendeleev
8. b) Covalent bond
9. c) Chlorine
10. b) Sublimation

Part IV. Biology

Question 1.
What is the basic unit of life?
  a) Cell
  b) Atom
  c) Molecule
  d) Organ

Question 2.
Who is known as the father of genetics?
  a) Charles Darwin
  b) Gregor Mendel
  c) James Watson
  d) Rosalind Franklin

Question 3.
What process do plants use to convert sunlight into food?
  a) Photosynthesis
  b) Respiration
  c) Fermentation
  d) Osmosis

Question 4.
Which molecule carries genetic information?
  a) RNA
  b) DNA
  c) Protein
  d) Lipid

Question 5.
What is the term for a group of organisms of the same species living in a given area?
 a) Community
 b) Population
 c) Ecosystem
 d) Biome

Question 6.
How many chromosomes do humans have in each cell?
 a) 23
 b) 46
 c) 92
 d) 44

Question 7.
What is the powerhouse of the cell, responsible for producing energy?
 a) Nucleus
 b) Mitochondria
 c) Ribosome
 d) Endoplasmic reticulum

Question 8.
Which type of cells have a nucleus?
 a) Prokaryotic cells
 b) Eukaryotic cells
 c) Both prokaryotic and eukaryotic cells
 d) Neither prokaryotic nor eukaryotic cells

Question 9.
What is the name of the process by which cells divide
to produce new cells?
    a) Meiosis
    b) Mitosis
    c) Fusion
    d) Fission

Question 10.
Which biome is characterized by low rainfall and extreme temperatures?
    a) Rainforest
    b) Tundra
    c) Desert
    d) Grassland

Answers

1. a) Cell
2. b) Gregor Mendel
3. a) Photosynthesis
4. b) DNA
5. b) Population
6. b) 46
7. b) Mitochondria
8. b) Eukaryotic cells
9. b) Mitosis
10. c) Desert

Part IV. Physics

Question 1.
What is the speed of light in a vacuum?
a) Approximately 3.0×10^8 meters per second
b) Approximately 1.5×10^8 meters per second
c) Approximately 2.2×10^7 meters per second
d) Approximately 5.0×10^8 meters per second

Question 2.
Who is known for the theory of relativity?
a) Isaac Newton
b) Albert Einstein
c) Niels Bohr
d) Max Planck

Question 3.
What particle is the quantum of electromagnetic field?
a) Electron
b) Proton
c) Photon
d) Neutron

Question 4.
What force is responsible for keeping planets in orbit around the Sun?
a) Electromagnetic force
b) Gravitational force
c) Strong nuclear force
d) Weak nuclear force

Question 5.
What is the principle that states the position and velocity of an object cannot both be measured exactly, at the same time, even in theory?
a) Newton's First Law
b) Conservation of Energy
c) Heisenberg's Uncertainty Principle
d) The Principle of Relativity

Question 6.
What unit is used to measure electric resistance?
a) Ohm
b) Volt
c) Ampere
d) Watt

Question 7.
Who discovered the electron?
a) J.J. Thomson
b) James Chadwick
c) Ernest Rutherford
d) Dmitri Mendeleev

Question 8.
Which of the following is a scalar quantity?
a) Velocity
b) Force
c) Mass
d) Acceleration

Question 9.

What phenomenon explains the bending of light rays when they pass from one medium into another?
 a) Reflection
 b) Refraction
 c) Diffraction
 d) Polarization

Question 10.

What is the first law of thermodynamics also known as?
 a) Law of Conservation of Momentum
 b) Law of Conservation of Mass
 c) Law of Conservation of Energy
 d) Law of Universal Gravitation

Answers

1. a) Approximately 3.0 x 10^8 meters per second
2. b) Albert Einstein
3. c) Photon
4. b) Gravitational force
5. c) Heisenberg's Uncertainty Principle
6. a) Ohm
7. a) J.J. Thomson
8. c) Mass
9. b) Refraction
10. c) Law of Conservation of Energy

Part V. Earth Sciences

Question 1.
What is the Earth's outermost layer called?
  a) Mantle
  b) Core
  c) Crust
  d) Lithosphere

Question 2.
What is the theory that describes the movement of Earth's plates?
  a) Evolutionary theory
  b) Plate tectonics
  c) Big Bang theory
  d) Quantum theory

Question 3.
What type of rock is formed from the cooling and solidification of magma or lava?
  a) Sedimentary
  b) Metamorphic
  c) Igneous
  d) Plutonic

Question 4.
What is the primary gas responsible for the greenhouse effect on Earth?
a) Oxygen
b) Nitrogen
c) Carbon dioxide
d) Methane

Question 5.
Which layer of the atmosphere is closest to Earth?
a) Stratosphere
b) Troposphere
c) Mesosphere
d) Thermosphere

Question 6.
What natural disaster is measured with the Richter scale?
a) Tornadoes
b) Earthquakes
c) Volcanic eruptions
d) Tsunamis

Question 7.
Which ocean is the largest?
a) Atlantic Ocean
b) Indian Ocean
c) Pacific Ocean
d) Arctic Ocean

Question 8.
What process describes the water cycle's movement from the surface to the atmosphere and back?
  a) Photosynthesis
  b) Respiration
  c) Evaporation
  d) Condensation

Question 9.
What is the name given to the hypothesis that the continents have moved over geologic time from one location to another?
  a) Plate shifting
  b) Continental drift
  c) Tectonic movement
  d) Earth migration

Question 10.
Which mineral is known as the hardest substance on Earth?
  a) Quartz
  b) Diamond
  c) Topaz
  d) Ruby

Answers

1. c) Crust
2. b) Plate tectonics
3. c) Igneous
4. c) Carbon dioxide
5. b) Troposphere
6. b) Earthquakes
7. c) Pacific Ocean
8. c) Evaporation
9. b) Continental drift
10. b) Diamond

Part VI. Environmental Sciences

Question 1.
What is the main cause of global warming?
    a) Decrease in solar radiation
    b) Increase in the Earth's orbit eccentricity
    c) Increase in greenhouse gas emissions
    d) Natural climate cycles

Question 2.
Which gas is most commonly associated with the depletion of the Earth's ozone layer?
    a) Carbon dioxide
    b) Chlorofluorocarbons (CFCs)
    c) Methane
    d) Nitrous oxide

Question 3.
What term describes the variety of life in the world or in a particular habitat or ecosystem?
    a) Ecology
    b) Biodiversity
    c) Conservation
    d) Biogeography

Question 4.
Which of the following is a renewable energy source?
    a) Coal
    b) Natural gas
    c) Solar energy
    d) Nuclear energy

Question 5.
What process involves the conversion of forested areas to non-forest land for agriculture, logging, or urban use?
a) Afforestation
b) Reforestation
c) Deforestation
d) Desertification

Question 6.
What is the largest source of freshwater on Earth?
a) Rivers
b) Lakes
c) Groundwater
d) Glaciers and ice caps

Question 7.
Which international agreement aims to reduce carbon emissions and combat global warming, signed in 2015?
a) Kyoto Protocol
b) Paris Agreement
c) Montreal Protocol
d) Stockholm Convention

Question 8.
What phenomenon occurs when pollutants cause a body of water to become overly enriched with minerals and nutrients, leading to excessive growth of algae?
a) Acidification
b) Eutrophication
c) Sedimentation
d) Salinization

Question 9.
Which species is often referred to as a "keystone species" because of its crucial role in maintaining the structure of an ecological community?
a) Honeybee
b) Grey wolf
c) Bengal tiger
d) Sea otter

Question 10.
What is the practice of managing the use of natural resources wisely so they can sustain future generations called?
a) Resource allocation
b) Sustainable development
c) Environmentalism
d) Conservationism

Answers

1. c) Increase in greenhouse gas emissions
2. b) Chlorofluorocarbons (CFCs)
3. b) Biodiversity
4. c) Solar energy
5. c) Deforestation
6. d) Glaciers and ice caps
7. b) Paris Agreement
8. b) Eutrophication
9. b) Grey wolf
10. b) Sustainable development

About The Author

Meet KJ Smith, a passionate amateur science enthusiast with an insatiable curiosity about the mysteries of the universe. While KJ's formal education may not be in science, their love for the subject knows no bounds.

From a young age, KJ was captivated by the wonders of the natural world, spending countless hours exploring backyard ecosystems and conducting homemade experiments. While their career may have taken them down a different path, KJ's fascination with science has remained a constant source of joy and inspiration.

"The Ultimate Science Trivia Book" is the brainchild of KJ's lifelong love affair with science. Fueled by their endless curiosity and eagerness to learn, KJ has meticulously compiled a collection of fascinating facts and mind-boggling trivia from across the scientific spectrum.

Whether it's pondering the intricacies of quantum mechanics or marveling at the wonders of the cosmos, KJ's enthusiasm for science shines through on every page. With a knack for making complex concepts accessible and engaging, KJ invites fellow science lovers of all levels to join them on a journey through the captivating world of scientific discovery.

When KJ isn't delving into the depths of the cosmos, you can find them exploring hiking trails, experimenting in the kitchen, or getting lost in a good book. With a boundless curiosity and a passion for sharing knowledge, KJ embodies the spirit of the amateur scientist, eager to uncover the secrets of the universe one trivia question at a time.